SAE EDGE™
RESEARCH REPORT

I0063012

Unsettled Issues Concerning the Opportunities and Challenges of eVTOL Applications during a Global Pandemic

Johnny T. Doo
International Vehicle Research, Inc.

EDGE DEVELOPMENT TEAM

Christopher M. Edens, MD, *US Air Force*

Michael R. Smith, *Bell Textron, Inc.*

Marilena D. Pavel, PhD, *Delft University of Technology*

Carl Dietrich, *Jump Aero, Inc.*

Ed De Reyes, *Sabrewing Aircraft Co.*

SAE
INTERNATIONAL®

Warrendale, Pennsylvania, USA

About the Publisher

SAE International® is a global association of more than 128,000 engineers and related technical experts in the aerospace, automotive, and commercial vehicle industries. Our core competencies are lifelong learning and voluntary consensus standards development. Visit sae.org

SAE EDGE™ Research Report Disclaimer

SAE EDGE™ Research Reports focus on topics that are dynamic, in which knowledge is incomplete, and which have yet to be standardized. They represent the collective wisdom of a group of experts and serve as a practical guide to the reader in understanding unsettled subject matter. They are not meant to provide a recommended practice or protocol. The experts have assembled as a community of practitioners to contribute and collectivize their thoughts and points of view; these are not the positions of the institutions or businesses with which they are affiliated, nor is one contributor's perspective advanced over other contributors. SAE EDGE™ Research Reports are the property of SAE International, and SAE alone is responsible for their content.

About This Publication

SAE EDGE™ Research Reports provide state-of-the-art and state-of-industry examinations of the most significant topics in mobility engineering. SAE EDGE™ contributors are experts from research, academia, and industry who have come together to explore and define the most critical advancements, challenges, and future direction in areas such as vehicle automation, unmanned aircraft, cybersecurity, advanced propulsion, advanced manufacturing, Internet of Things, and connectivity.

Related Resources

SAE MOBILUS® Automated & Connected Knowledge Hub
https://saemobilus.sae.org/automated-connected/

SAE Team

Frank Menchaca, Chief Growth Officer

Michael Thompson, Director of Standards, Information and Research Publications

Monica Nogueira, Director of Content Acquisition

Beth Ellen Dibeler, Product Manager

William Kucinski, Managing Technical Editor

EPR2020022
ISSN 2640-3536
e-ISSN 2640-3544
ISBN 978-1-4686-0252-4

To purchase bulk quantities, please contact: SAE Customer Service

E-mail: CustomerService@sae.org
Phone: 877-606-7323 (*inside USA and Canada*)
          +1-724-776-4970 (*outside USA*)
Fax: +1-724-776-0790

https://www.sae.org/publications/edge-research-reports

About the Editor

Mr. Johnny T. Doo is the President of International Vehicle Research, Inc., focusing on innovative flight vehicle and electric vertical take-off and landing (eVTOL) technologies and unmanned aerial vehicle (UAV) and wing-in-ground vehicle development and applications. He is the Lead of the NASA Transformative Vertical Flight Working Group on Public Services, teaming with over 100 industry leaders and experts to develop the roadmap and use cases for disaster relief, humanitarian aid, search and rescue, police and firefighting, medical transport, and military eVTOL applications. He also supports aviation industry growth by providing strategic consulting, including business planning, product development, and market strategy to selected global clients.

He has over 30 years of experience in manned and unmanned aviation products, with many years of executive-level responsibilities of aviation product development, program management, supply chain development, as well as marketing and business development. His technical and program experiences range from high-performance piston aircraft, personal and business jets, regional aircraft, new-generation aircraft engines to industrial UAVs. His expertise includes aeronautical engineering, product design and testing, advanced technology, aircraft certification, and project management, with particular focuses on design-to-cost and design-for-production aspects.

He was the Executive Vice President of Advanced Technology, Executive Vice President of Marketing & Business Development, and Senior Vice President of Engineering & Product Integrity at Continental Motors Group. He not only managed the product line expansion of the Avgas and Jet-A diesel aviation engines, but also developed multiple new market segments, achieving significant growth of the global original equipment manufacturer customer base and sales revenue. He was also president of a flight training subsidiary with an innovative training model and was responsible for the maintenance, repair, and overhaul business unit, quadrupling business growth over three years.

Prior to Continental Motors, he was the Vice President at Diamond Aircraft in Canada, with full responsibility for the new-generation personal jet program. He built a new program team from the ground up and led the successful development of the all-composite, single-engine personal jet in just three years. At the same time, he established the design-built-test capabilities, negotiated with all the key suppliers, and managed the new supply chain to support the development and production operations. His hands-on experiences with the high-performance business jet at Sino Swearingen Aircraft, regional aircraft at Fairchild Aircraft, and new-generation piston aircraft at Mooney Aircraft formed a broad technical and project management knowledge base.

He has a Bachelor's degree in Mechanical Engineering from Chung Yuan Christian University and a Master's degree in Aerospace Engineering from the University of Colorado. He is the lead author of "NASA Electric Vertical Takeoff and Landing (eVTOL) Aircraft Technology for Public Services" white paper, co-author of the book *WIG Craft and Ekranoplan—Ground Effect Craft Technology*, and is the GoFly Master Lecturer on "eVTOL for Public Services—Design & Applications."

contents

Atosan/Shutterstock.com

Unsettled Issues Concerning the Opportunities and Challenges of eVTOL Applications during a Global Pandemic

Abstract

Recent advancements of electric vertical takeoff and landing (eVTOL) aircraft have generated significant interest within and beyond the traditional aviation industry, and many new and novel applications have been identified and under development. The COVID-19 crisis has highlighted the challenges of managing a global pandemic response due to the difference in regional and local resources, culture, and political systems. Although there may not be a uniform crisis management strategy that the world can agree on, we can leverage a new generation of vertical flight vehicles to make a difference if (or when) such a global epidemic strikes again. One of the key challenges realized in the early stage of the COVID-19 outbreak is the ability to allocate and distribute limited and critical medical resources, including equipment, supplies, medical personnel, and first responders to the hot spots when and where they may be needed. The on-demand logistics capabilities could be enhanced by the availability of new-generation eVTOL aircraft and their forthcoming autonomous operation. The ability to land and takeoff at many unconventional locations makes eVTOL aircraft valuable assists for complementing and enhancing on-demand logistic needs. To make such operations truly productive is not easy; there are reassociated challenges that needed to be addressed to enable the benefit of such a system. In addition to having a large enough eVTOL fleet near a location that can be available for the pandemic responses, the support system and the ability to deploy and reposition the fleet dynamically with supporting infrastructures are also required. Based on the rapid development progress of eVTOL, it is envisioned that those challenges can be addressed soon.

JOHNNY T. DOO
International Vehicle Research, Inc.

Edge Development Team
Christopher M. Edens, MD, *US Air Force*
Michael R. Smith, *Bell Textron, Inc.*
Marilena D. Pavel, PhD, *Delft University of Technology*
Carl Dietrich, *Jump Aero, Inc.*
Ed De Reyes, *Sabrewing Aircraft Co.*

ISSN 2640-3536

Introduction

State of the Industry

The electric vertical take-off and landing (eVTOL) industry is developing and evolving rapidly. It is not just about creating a new vertical flight vehicle but developing a complete ecosystem covering multiaspect technologies, vehicle development and certification, autonomous and simplified vehicle operation (SVO), air space management and integration, high-volume and cost-effective production, infrastructures, and community engagement—as well as market development and investments.

Significant progress has been made in the last few years, with over 300 eVTOL aircraft designs currently in various stages of development [1]. Multiple programs and efforts have also been initiated or are in progress that are organized or supported by government agencies, including NASA Advanced Air Mobility (AAM), Air Force Agility Prime, NASA Transformative Vertical Flight Working Groups, and many others [2, 3, 4]. The Federal Aviation Administration (FAA) and the European Union Aviation Safety Agency have also been working toward a viable certification basis and means of compliance to enable the industry to move forward with accelerating speed.

As a new industry, the main challenges and opportunities in the eVTOL evolution involve vehicle performance and capabilities, new and novel applications, as well as the infrastructure and support challenges that come with them. In addition to urban air mobility (UAM)/AAM opportunities, another type of use has received increasing attention recently—public service applications, which include medical transport, cargo logistics, search and rescue, disaster relief, humanitarian aid, and military operations. This report discusses the unsettled issues related to the opportunities and critical challenges of eVTOL applications in a global pandemic.

Unsettled Domains in eVTOL Applications in a Global Pandemic

The world has faced unexpected challenges in effectively responding to the COVID-19 pandemic. The novel virus, which is named "SARS-CoV-2" (Severe Acute Respiratory Syndrome [SARS]-Coronavirus [CoV]), results in the disease that has been named "Coronavirus Disease 2019" (COVID-19). The virus spread like wildfire across the globe, with more than 39 million cases being counted from January to mid-October 2020 [5]. The measures that governments across the world have taken, especially between December 2019 and June 2020, to contain the spread of coronavirus disease were massive and unprecedented [6]. Never in the history of humanity have such drastic lockdowns occurred at the nationwide levels, motivated by shocking images of

overwhelmed hospitals and extension of the disease. As a result of these measures, life came to an almost complete standstill in many regions in the world (e.g., the lockdown strategy deployed in and around Wuhan region in China involved 56 million people). The problems were not only global but also regional and local, with corresponding differences in strategies and practices.

One critical issue identified in the COVID-19 situation was the importance of logistics to mitigate the negative impact of the pandemic in terms of humanitarian response, supply chains, and economic costs [7]. In this sense, eVTOL aircraft have the ability to make a difference in the battle against the epidemic. Due to the global nature of the pandemic, the logistical need for medical supplies and equipment to counter the virus has become one of the key challenges. No country or region was sufficiently prepared with basic and specialized medical supplies, such as personal protective equipment (PPE) and respirators, to allow every hospital to be fully stocked and equipped with two to three months of inventory, even as the production of much-needed goods was rapidly ramping up. In China, in February 2020, EHang demonstrated the use of its eVTOL aircraft for transport of medical supplies and personnel as part of China's response to the coronavirus outbreak [8].

As the world faces shortages of crucial equipment and supplies, it is feasible—and indeed imperative—to adopt a much more dynamic logistics system to deliver goods and personnel on-demand to the needed locations. In the literature for general service logistics, there is an increased interest in transforming "static service operations" to "bring-service-near-your-home" mobile service operations [9]. For eVTOL, this capability means strategically positioning equipment and assets at specific depot locations, allowing limited supplies to support critical hospitals or satellite sites without excessive inventory at each location. In addition, dynamic allocation and transportation of medical equipment, as well as food supplies and agricultural products in a timely manner, can be accomplished through a combination of medium-range and short-range eVTOL aircraft when the distribution or processing network is impacted or disrupted (Figure 1).

Benefits of future eVTOL use during a pandemic include the following:

- On-demand distribution to dynamically transport medical supplies and equipment from warehouses to hospitals

- Quick deployment (high cycle) to reallocate/repurpose existing air taxis (anticipated near-future capability) or public service fleets

- VTOL on streets/parking lots/rooftops/schoolyards to enable on-demand flexible operations

- Autonomous/semiautonomous transport for patients to reduce contact and contamination

- Medical and support personnel transportation to and from needed sites

- Availability of small local general aviation (GA) airports for potential use as logistics hubs

FIGURE 1. Most eVTOL are being developed to transport two to four passengers up to 60 miles at airspeeds around 135 knots. eVTOL evolution and rapid development can greatly enhance responsiveness during future pandemics.

Unsettled Issues

Achieving a future eVTOL environment that could contribute to global pandemic relief and support efforts is likely a reality in the near future; however, significant challenges must first be addressed:

- **Government embracement**: Federal, state, and local agencies all need to embrace the use of eVTOL aircraft in such emergency conditions.

- **Community acceptance and support**: First responders, medical personnel, logistic operators, and the general public also need to accept the use of eVTOL aircraft for emergency needs.

- **Vehicle technologies and capabilities**: Qualities such as performance and adaptability for pandemic missions need to be developed.

- **Vehicle operations and maintenance**: SVO, autonomy, and simplified maintenance qualities are critical for effective eVTOL aircraft deployment.

- **Flight management and airspace control network**: The variety of missions for which an eVTOL aircraft might be needed requires airspace managers to rearrange with short notice.

- **On-demand medical equipment retrofit and integration/pilot protection**: The need for integrated isolation chambers on an eVTOL aircraft—as well as pilot and operator protection against viral contamination (including ground personnel)—is vital.

- **Availability of affordable units in mass quantity globally**: Specially equipped eVTOL aircraft need to be developed economically in order to be deployed and allocated in an effective manner.

- **Dynamic deployment infrastructure and logistic support**: Dynamic allocation of usable eVTOL aircraft must account for fast-changing missions and diverse geographic deployment.

eVTOL Technology: Past, Present, and Future

eVTOL Systems Currently under Development

eVTOL systems currently under development vary widely, from microscale unmanned aircraft systems (UAS) weighing less than 4.4 lb to small UAS (sUAS) weighing up to 55 lb, to the people-moving air taxis to be type certified in the future [10].* This remarkable breadth of eVTOL flight options enables an equally wide range of practical applications that span a broad spectrum of concept of operations (ConOps).

While it is beyond the scope of this SAE EDGE Research Report to comprehensively list or review all eVTOL aircraft currently in development, we will mention one family of air vehicles being developed by Bell Textron Inc. to illustrate how these can be used to complement global COVID-19 pandemic response capabilities. The Autonomous Pod Transport (APT) family of unmanned aerial vehicles (UAVs) currently consists of the APT20 and APT70 shown in Figures 2 through 4.

---

* In NPRM 84 FR 3865, the FAA's Micro UAS Aviation Rulemaking Committee (ARC) recommended four categories of operations of small UAS over people: ARC Category 1 sUAS weighing ≤0.55 lb would have no restrictions; sUAS in ARC Categories 2, 3, and 4 would be required to not exceed specified impact energy thresholds for each category with scaled up performance-based standards and operational restrictions for each successive category to mitigate the increased risks.

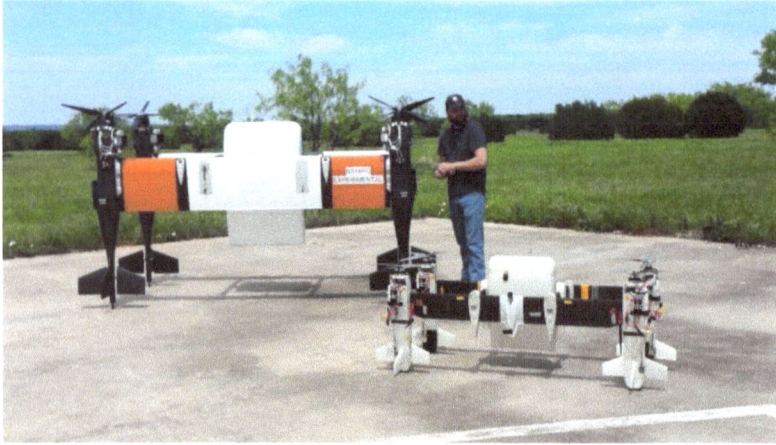

FIGURE 2. APT70 (left) and APT20 (right) UAS.

Reprinted with permission. © Bell.

FIGURE 3. APT70 UAS during early dawn take-off.

Reprinted with permission. © Bell.

FIGURE 4. APT70 in cruise flight.

Reprinted with permission. © Bell.

With a maximum gross weight just under 55 lb and operating as a *14 CFR Part 107* UAS, the APT20 carries 20 lb of payload up to 18 miles at a cruise speed of 60 to 80 mph. At 320 lb maximum gross weight, the APT70 carries 70 lb of payload up to 35 miles with a similar cruise speed. While the APT70 has demonstrated airborne detect and avoid capabilities, all UAS must be able to comply with *14 CFR § 91.113(b)* for integration of manned and unmanned air vehicles in the National Airspace System:

> *14 CFR § 91.113(b) Right of Way Rules: General. When weather conditions permit, regardless of whether an operation is conducted under instrument flight rules or visual flight rules, vigilance shall be maintained by each person operating an aircraft so as to see and avoid other aircraft. When a rule of this section gives another aircraft the right-of-way, the pilot shall give way to that aircraft and may not pass over, under, or ahead of it unless well clear. Amdt. 91-282*

Potential Ways to Use eVTOL to Complement Global Pandemic Response Capabilities

With the expansive growth of eVTOL applications, UAS registered in the US exceeding 1.7 million, and the immediate need to enhance public safety from and during the global COVID-19 pandemic, UAS utilization presents a prodigious and timely opportunity [11]. These airborne systems can be adapted to provide thermal sensing, crowd monitoring and dispersal, and transportation of critical medical supplies.

Thermal sensing from an overhead vantage point can be used to identify persons with elevated surface skin temperatures at or above 100.4° Fahrenheit/30° Celcius. This can be accomplished using telethermographic systems to identify potential infection sources and direct them away from crowds and to medical triage facilities.

From the same overhead perspective, crowds not observing social distancing protocols during peak episodes of disease transmission can be detected. Once detected, the UAS can be used to broadcast a warning to safely separate in order to mitigate the adverse public spread of airborne communicable diseases.

Another example where eVTOL can complement the battle against a pandemic can be related to the creation of a flexible and "very short notice" preparedness for a coronavirus testing system, ensuring, for example, swab tests logistics.

Finally, the transport of critical medical supplies can be rapidly and broadly utilized for optimized distribution in a timely manner.

Technology Readiness Level

Using NASA's Technology Readiness Level (TRL) definitions, only UAS systems at TRL 7 or higher are ready for implementation in the response against COVID-19 [12]. However, continued development is needed to push the TRL of these systems to TRL 8 to be fully certified for safe, integrated operations in the National Airspace System.

Future applications include safe and rapid transport of quarantine critical patients once the people-carrying eVTOL aircraft, such as the air taxi, have been matured to TRL 8. The capabilities of leveraging autonomous and semi-autonomous cargo transportation are likely to be approved and accepted for low-density population areas first, then expand into full-scale deployable network.

The NASA AAM project focuses on enabling emerging aviation markets that will provide substantial benefit to the US public and industry [13]. Starting with UAM, the AAM project will enable these markets by being a community catalyst and developing and validating system-level concepts and solutions both within AAM and in coordination with other NASA Aeronautics Research Mission Directorate projects. These groups already play a role in enabling UAM technology. The integrated approach is expected to accelerate the eVTOL industry TRL level advancement and enable product and system commercial maturity in a shorter time frame.

Military interest in eVTOL technology and systems can accelerate the development and implementation, and the US Air Force Agility Prime initiative is a good example [14]. Working with industry partners will help turn what was once thought to be the impossible into a reality, benefiting both the military and civilian sectors, decrease risk in testing, and the amount of time to commercialization. With a projection as early as 2023, Agility Prime is leveraging the strong military demand on the emerging market—making it clear to investors, regulators, and innovators in the tech space that the military wants to start purchasing these innovative vehicles as soon as they can.

Key challenges and development needed to advance the eVTOL industry, especially for global pandemic support, will be further discussed in the latter part of this report:

- Vehicle-specific capabilities
- Autonomy and air space management
- Operations, logistics, and support
- Unconventional medical transport mode evaluation
- Funding and acceptance

Current Medical and Logistic Transport Operations—Ground and Air

Current medical and logistic transport operations can be broadly divided into emergency and nonemergency domains, with both domains making use of ground and air transportation platforms. Various emergency medical transport platforms provide 640,000 critical care patient transfers annually in the US, with 230,000 transports via ground, 300,000 via helicopter, and 40,000 via critical care fixed-wing platforms [15]. Medical transfers primarily involve transportation of a patient from the scene of an incident to an initial hospital or point of care (primary transfer) or from a nonspecialized hospital to a higher level of care (secondary transfer).

Ground-Based Transport and Capabilities

Emergency Ground Transportation Ground emergency medical services (GEMS) consist of emergency medical services (EMS) providers and ground ambulances. While the ubiquitous ambulance may be familiar to all, many outside the medical field may not be aware of the differences in levels of training and certification for the various EMS providers and their differing scopes of practice, such as the emergency medical responder or "first responder," the emergency medical technician (EMT), advanced EMT, and paramedic. Similarly, ground transportation ambulances are divided into basic life support ambulances and advanced life support ambulances—the latter of which is staffed by paramedics.

Nonemergency Ground Transportation Many additional transportation modalities are used for nonemergency patient movement, such as the non-EMS ambulance. Vans and other vehicle-based transportation are also used for long-distance

patient transportation. For local movement, wheelchairs and motorized scooters are used by patients and transportation services, typically for "last mile" movement after other modalities have been used. Finally, courier services are an important link in the transportation of human blood, organs, and other biological matter to and from hospitals and other facilities.

Air Transportation— Helicopters and Fixed Wings

Emergency Air Transportation Helicopter air ambulances (HAAs, Figure 5), previously referred to as helicopter emergency medical services (or "HEMS"), represent the most common emergency air transportation platform and are typically staffed with highly skilled EMS providers and flight nurses with more advanced training and greater scope of practice than GEMS [16]. In some states or jurisdictions, police helicopters conduct the initial patient rescue from an accident scene while private air ambulance companies provide interfacility transfer and backup to police helicopters. In the US, there were 75 air ambulance companies operating 1,515 helicopters as of 2014.

Fixed-wing platforms are much less commonly used for emergency air transportation; however, they do fill a role in longer-distance transfer and medical evacuation (MEDEVAC). One of the most common applications in this category is military MEDEVAC, pioneered by the US military medical system during the Korean War. While initial medical care was centered at far-forward medical facilities (during the Korean, Vietnam, and the Gulf War), by 2005, the model had evolved to rapid stabilization and transportation of critically injured patients out of the theater for definitive care. Initial

FIGURE 5. An emergency helicopter air ambulance (HAA).

Hypervision Creative/Shutterstock.com

FIGURE 6. **Fixed-wing medical air transport of the International Committee of the Red Cross.**

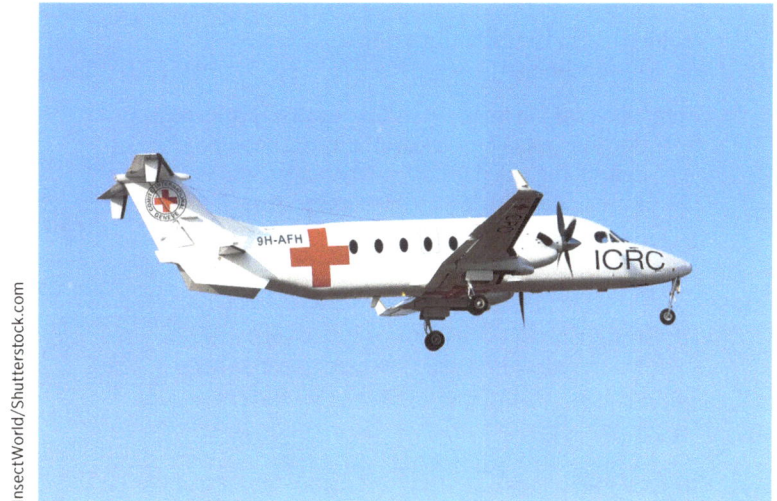

InsectWorld/Shutterstock.com

rotary-wing MEDEVAC aircraft were rapidly followed by fixed-wing aircraft, which had a <0.02% en route mortality rate and 2.1% subsequent 30-day mortality rate. Central to the success of this model was the three-person critical care air transport team that self-carried required equipment for transport and critical care of battle-injured patients [17].

Nonemergency Air Transportation Both rotary- and fixed-wing aircraft are occasionally used for nonemergency medical transport, which includes moving stabilized patients to a higher level of care or for medical evacuation in disaster situations. Helicopters are occasionally used for prescheduled, routine transfers between hospitals, most often from rural to metropolitan hospitals. However, for longer-distance missions, the speed, safety, space, weight limit, and pressurized environment advantages of fixed-wing platforms (Figure 6) become apparent. Military and civilian fixed-wing aircraft from various nations have played significant roles in recent large-scale disaster response, including after terrorist attacks in Bali (2002), Mombasa (2002), and Madrid (2004); during Middle East conflict (2003 to 2006); and ahead of the Indian Ocean tsunami (2004) [18].

Interfacility Transfer Interfacility transfer (IFT) refers to the movement of a patient from one healthcare facility to another, primarily in regions with large rural populations (e.g., moving a patient from a lower to a higher level of care). In the US, as in many other countries, rural community hospitals typically offer a lower level of medical, surgical, and trauma care, often necessitating transfer to a larger, often urban-based facility. In an example from Switzerland, 79% of helicopter-based IFTs were from regional hospitals to a university hospital for upgrading the level of care [19].

Both ground and air transportation platforms are used for IFT, with ground-based ambulances and helicopters being the most common. For trauma patients requiring IFT to a tertiary trauma center, the most influential factor was distance, with helicopters preferred for IFT greater than 45 miles. However, many additional factors were inconsistently associated with the preference of helicopter over GEMS, including the type and severity of the injury, shock, traumatic brain injury, and arrival via GEMS at the initial hospital. Additionally, the lack of availability of rural GEMS, which have been struggling to stay in business, has led to other services (particularly HAAs) covering more rural areas [20]. Data suggests that rapid, initial transfer of severely injured (and potentially ill) patients directly to a high-level healthcare facility for definitive treatment may be critical to preserving life, especially for rural patients [7].

Factors Affecting Transportation Modality

Many independent and interdependent factors affect the decision of using ground versus air transportation assets for both emergency and nonemergency patient transfer. The most critical factors contributing to this decision in an emergency include the following [21]:

- Referral factors: The skills of the initial on-scene responder in generation of the initial diagnosis and intervention plan, as well as the proximity and familiarity with the transportation resources available, greatly affect the transportation decision.

- Patient factors: Key considerations include the disease or injury, response to initial treatment, the need for ongoing interventions, and the perceived urgency of the transfer.

- Receiving hospital: The ability of the receiving hospital to accept and treat patients include both bed capacity and the medical treatment capability required (e.g., trauma surgery, intensive care, cardiac care, etc.).

- Transportation factors: GEMS is negatively affected by traffic, distance, and occasionally availability (e.g., decreasing rural GEMS availability), while HAAs are adversely affected by weather, the availability of landing zones at each destination, and availability (as many regions may have only one or two HAA units available).

- Logistical factors: Logistical considerations include crew work hours and fatigue management, other demands on the transportation teams, and cost.

Key Advantages and Disadvantages of Ground versus Air Transportation GEMS may be preferable for moderate distances (less than 45 minutes from trauma center), as GEMS response times may be less than an HAA. This is primarily driven by the greater time from initial dispatch, engine startup, take-off, and landing required for an HAA [16]. In particular, in the initial response, GEMS assets are typically much closer from the dispatch location to the patient than an HAA, resulting in a shorter mean dispatch-to-scene arrival time: five minutes (plus or minus one minute) with GEMS versus 22 minutes (plus or minus three minutes) with an HAA [22].

HAAs offers faster travel time than GEMS, but this travel time advantage is primarily seen at greater distances from the medical facility or when peak traffic is negatively affecting GEMS. However, adverse weather negatively impacts HAAs, making them the less preferred modality (Table 1).

Medical Transport and Logistic Challenges

The COVID-19 pandemic illustrated the ability of infectious diseases to spread quickly both within and between countries (ultimately, globally). This is particularly true of infectious respiratory viral diseases, including two prior coronavirus pandemics—Severe Acute Respiratory Syndrome or "SARS" and Middle East Respiratory Syndrome or "MERS"—as well as seasonal and epidemic/pandemic influenza (e.g., H1N1 "swine flu") in this century alone. While other viral and bacterial infectious diseases have the potential for epidemic or pandemic levels of spread, respiratory viral infections (particularly influenza) have represented the vast majority of the recent epidemic and pandemic outbreaks [23]. Accordingly, medical transportation should focus on the challenges posed by a prototypical respiratory viral pandemic.

While medical transportation of patients can present challenges in everyday rescue or transfer situations, movement of patients and medical supplies during a pandemic adds significant complexity with many additional and magnified challenges. Early in a pandemic, the number of new patient cases is low relative to medical transportation assets—with patient factors designated as the primary challenge. However, as case numbers rapidly or exponentially increase, patient demand may far exceed transportation asset availability—leading to subsequent population-level challenges. Finally, throughout a pandemic, the movement of tests, medical supplies, and equipment presents significant logistical challenges.

Emergency Patient Transportation Challenges

Transportation of individual patients in a pandemic presents significant risks to both the patient and the ambulance personnel. While exposure to the infectious agent is a key risk posed by the patient to the medical personnel (common to both ground and air transportation), the transportation platform—particularly air transportation—can pose significant risks to a critically ill patient. This section explores these facets of transportation challenges in a pandemic on a microscale at the patient-ambulance level.

TABLE 1. Key advantages and disadvantages of ground EMS, helicopter EMS, and near-future eVTOL EMS platforms.

	Ground EMS	Helicopter air ambulance	eVTOL EMS (anticipated)
Advantages	Low cost of operations	High transportation speed	Low cost of operation
	Large network (urban and suburban)	Potential mortality benefit for critical patients	High transportation speed
	Short dispatch time		Short dispatch time
	Efficient for short distances		Efficient for short to medium distances
			Distributed network
			Smart command and control
Disadvantages	Declining rural network	High cost of operation	Novel platform
	Slow at longer distances	Safety concerns	Requires certification and acceptance
	Limited by traffic	Limited network	Requires vertiport infrastructure
		Longer dispatch time	
		Weather-dependent	

FIGURE 7. The patient isolation unit (PIU) for highly infectious patient transports in pressurized cabins, developed by Swiss Air-Ambulance Rega.

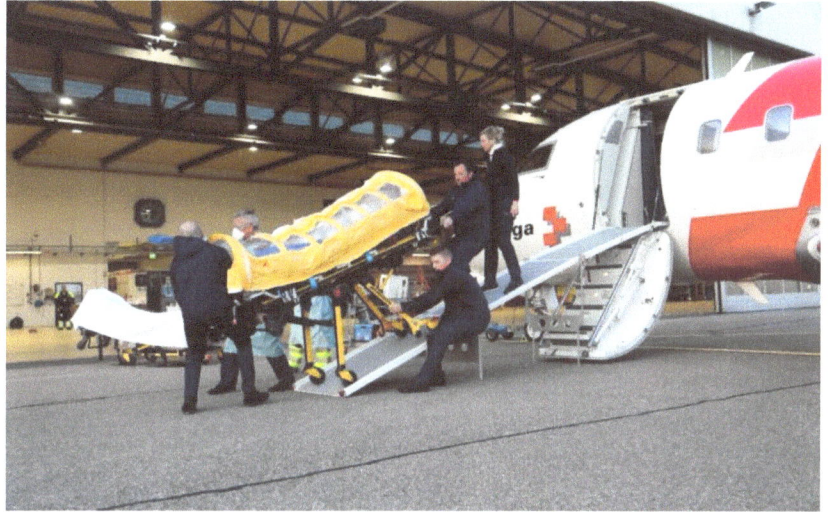

Reprinted with permission. © Rega

Risks to the Transportation System Respiratory viral infections are usually associated with coughing, which can easily spread infectious viral particles. The recent COVID-19 pandemic highlighted this risk to healthcare personnel from aerosolized viral particles, particularly in critically ill patients in respiratory distress requiring intubation and ventilatory support. While PPE is widely used by EMS transportation personnel, suboptimal real-world use and PPE shortages can decrease the effectiveness and allow for infection in healthcare workers. Accordingly, various barriers or patient isolation units (PIUs) were employed to protect ambulance crews, although the use of these systems requires additional ambulance space and training for appropriate use (Figure 7). Additionally, research prior to the COVID-19 pandemic highlighted the concerns with the prehospital EMS workforce's preparation for a respiratory viral pandemic, including deficits in knowledge, confidence, and willingness to work in a pandemic situation [24].

Bell, with support from Alpine Aerotech, has developed a crew barrier for the Royal Canadian Air Force CH-146 Griffon helicopter fleet to protect crews during the COVID-19 pandemic (Figures 8 and 9). The isolation system, which takes only minutes to install, reduces the risk of transmitting viral and bacterial diseases to flight crews without compromising the operational capability of the helicopter. In addition, a disinfectant solution has been deployed and used to further protect crew members. This barrier system is available for the Huey family of helicopters, and a similar system is being approved for the 206-series, including the Bell 407 helicopter.

FIGURE 8. Canadian Forces CH-146 Griffon.

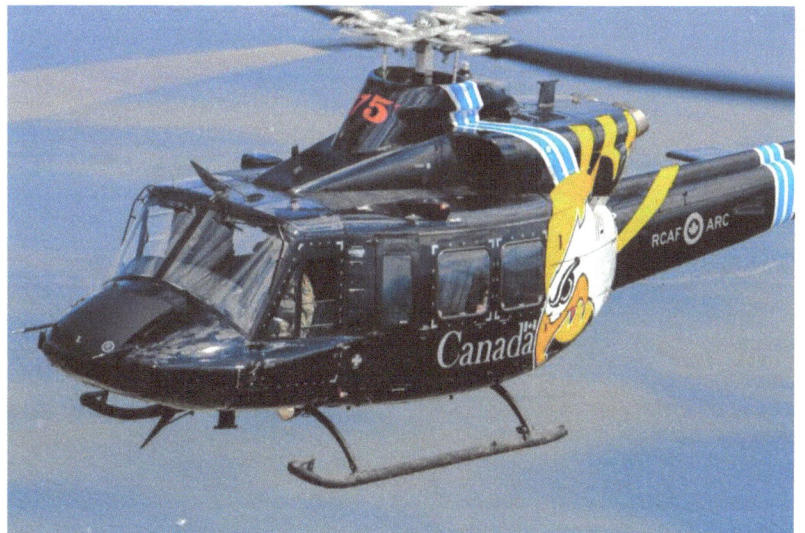

Reprinted with permission. © Bell

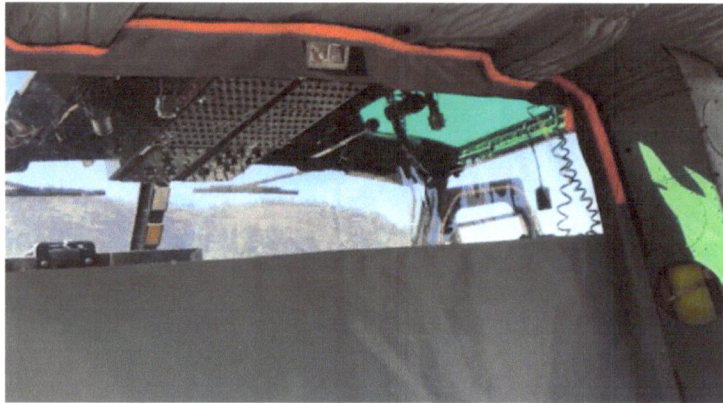

FIGURE 9. Crew Barrier for CH-146 Griffon.

Reprinted with permission. © Bell

Ground Patient Transportation in a Pandemic

During the COVID-19 pandemic, many countries (including most European countries) preferred or required ground-based transportation over air transportation of confirmed cases [25]. However, there are several key limitations of GEMS when filling this role. First, GEMS crews (particularly basic life support ambulance crews) typically are not as highly trained as air ambulance crews; they may not be sufficiently trained or equipped to respond to and transfer critically ill patients [26]. Additionally, GEMS outside major cities and in rural areas may be limited or unavailable due to the decline of ambulance services [20]. Finally, in dense urban areas (which are usually affected early in viral pandemics), ground ambulances are negatively affected by traffic and distance, reducing the speed of patient transfer, which may affect survival for critically ill patients and increase exposure time for the ambulance crew [15].

Air Patient Transportation in a Pandemic

Air transportation (primarily via helicopter) offers advantages over GEMS, including transportation speed and more advanced training of flight EMS crews. However, helicopter transportation of patients in a pandemic faces several major challenges, primarily the aeromedical environment, high costs, and significant safety risks.

The aeromedical environment encountered in air ambulances presents multiple challenges to critically ill patients, including noise, vibration, turbulence, thermal stresses, acceleration and deceleration, hypoxia, and pressure changes. Among these, hypoxia, thermal stresses, and pressure changes may represent particular challenges in viral pandemic patients in respiratory distress. In addition, the limited space available in helicopters may present a significant challenge for EMS to provide care for these patients, who may have intravenous lines placed and require oxygen via mechanical ventilation. Finally, as discussed, the use of patient barriers or isolation devices may further complicate the challenges posed by the aeromedical environment on patient care as well as emergency evacuation.

Second, HAAs represent a significant safety risk due to mishaps. Piloting an HAA is one of the most dangerous occupations in the US, with an accident rate exceeding 100 deaths per 100,000 employees (compared to 21 deaths per 100,000 police officers) [27]. A five-year review of the National Transportation Safety Board (NTSB) accident synopses database (1997-2001) found 47 accident files with 40 fatalities and 36 injuries, with 70% of incidents due to pilot error [28]. In a subsequent NTSB review (2011-2013) of 28 HAA accidents and incidents with 16 fatalities, HAAs had a two-fold higher crash rate versus other air taxis and a six-fold higher fatal accident rate per transport than GEMS [29]. The greatest risk factors included adverse weather, pilot disorientation, and lack of basic or advanced safety equipment (e.g., autopilot, terrain awareness and warning systems, night vision goggles, and flight data recorders).

A final significant challenge for HAAs is cost. Compared to GEMS, HAAs face tenfold higher hourly operating costs due to higher maintenance, fuel, and personnel costs, with the cost of airfare alone at $12,000-$25,000 per patient transported in the United States [22]. Notably, one in four patients transferred by HAA in the United States is discharged within 24 hours, indicating overuse contributing to higher healthcare system costs [19]. In the COVID-19 pandemic, patients who developed severe disease frequently experienced rapid deterioration, further complicating efforts by first responders to triage patients into high- or low-risk categories for appropriate methods of transfer. Thus, controlling costs through the reduction of overtriage and overuse of HAA transport may be quite difficult in a real-world, rapidly evolving pandemic situation, while other transportation modalities with lower operating costs (including eVTOL EMS) may represent the most viable pathway to cost-effectiveness.

Very limited data is available on the use of fixed-wing aircraft in the transportation of pandemic patients, largely due to the similarly limited role for longer-range transportation of pandemic patients. Fixed-wing air ambulances share some of the limitations and risks with helicopters, including the aeromedical environment and costs (varying depending on platform). However, compared to HAAs, fixed-wing air ambulances typically have advantages in pressurization, temperature control, and greater working space. Key limitations are the requirement for patient transfer to and from an airfield, requiring additional time and medical personnel for multimodal transportation [25].

In summary, improving outcomes on a micro-level for pandemic patients must address three core areas of concern. First, improving health outcomes may be targeted by both improvements in total patient transfer efficiency and the potential inclusion of skilled healthcare personnel in the transportation process. Second, the safety of ambulances (primarily air ambulances) must be significantly improved to include the capability to operate in all-weather, night, and low-level or mountainous terrain where current HAAs struggle. Finally, cost-effectiveness must be addressed, with the greatest potential gains possible through significant reductions in operating costs with novel eVTOL platforms.

Transportation System Challenges in Major Medical Response Efforts

Medical Response Challenges in Major Disasters
As cases rise during a pandemic, systems-level challenges quickly emerge as significant factors. In major medical response situations (including mass critical-care situations and mass-casualty incidents), the scale of response required can quickly overwhelm transportation assets. The challenge of major medical response to disasters has been repeatedly illustrated by major earthquakes, including the Kumamoto earthquake in Japan (2016) and the Kermanshah earthquake in Iran (2017). Two key challenges are shared between these large-scale disasters and a potential worldwide viral pandemic: the need for specialized or rapidly adaptable platforms with crews trained for a major medical response and maintaining effective command, control, and communications through a centralized coordinating center.

Specialized or Adaptable Platforms
Major medical response situations demand platforms and appropriately trained crews capable of effectively responding to a significant surge in demand for transportation resources. Specialized EMS transportation assets may not be available at the pandemic scale, requiring adaptation of nonspecialized transportation assets. In several prior major medical disaster responses, nonspecialized fire department and military helicopters were largely unable to respond effectively when medical helicopters were unable to meet demand [30]. A large-scale pandemic response requires the ability to easily, quickly, and safely load and unload patients into and from transport vehicles. In a prototypical respiratory viral pandemic, transport vehicles must be able to accommodate nonambulatory, litter-bound, and potentially critical care patients, including the medical equipment required for successful ongoing care and transportation of these patients. These equipment requirements may include ventilator support, oxygen, stations for hanging intravenous infusion medications and fluids, and power for

medical devices. At a minimum, PPE and potentially portable PIUs must be available and able to be used on transportation platforms. While nonspecialized transportation assets may be available, consideration must be given to the necessary adaptations required for use in pandemic response, including the training of transportation crews.

The most significant barrier for permanent, dedicated medical transportation assets is cost, especially considering the infrequent to rare use for disaster or pandemic patient transportation. Accordingly, in addition to the adaptation of nonmedical platforms, modular configurations have demonstrated successful and cost-effective use in prior major medical responses. As an example, the Swedish National Air Medevac System used a Boeing 737-800 in a modular configuration that could be converted for medical use in less than six hours. A modified version of this system was successfully used in the 2004 Indian Ocean tsunami and 2008 Mumbai terrorist attacks. Modular medical systems may allow for the use of civil or military transportation assets, including fixed-wing aircraft and larger helicopters [18].

Centralized Coordinating Center for Command, Control, and Communications
The lack of an effective integrated management system represents one of the most common failure points in prior major disaster response efforts. In the 2017 Kermanshah earthquake in Iran, the lack of systemic coordination and unified command was the most important factor compromising an effective response. Critical communication links were missing between air and ground EMS units and between helicopter operators from various medical, military, and rescue organizations [31]. In contrast, the Japanese emergency medical information system was developed to share critical disaster-related information (including hospital demand versus use and medical assistance team status). The system featured real-time, internet-based monitoring with command and control. While this system promoted a largely successful response to the 2016 Kumamoto earthquake in Japan, several minor issues surfaced including the need to optimize flow of information on medical demand, real-time information sharing, coordination with air traffic control, and wireless communications procedures. A more difficult issue to solve was how to prioritize patients for transportation and match them to the best mode of transportation [30]. Ultimately, the centralized coordinating center in a pandemic must be resilient and able to organize and communicate a large volume of information to effectively command and control the medical transportation system through dynamic allocation of all available resources on regional, national, and global levels.

Medical Supply and Equipment Logistical Challenges

The success of operations in a peacetime major medical response (as in wartime) is critically dependent on medical logistics, including facilities, personnel, supplies, equipment, and operating procedures for supply and resupply. Medical logistics during a major disaster faces multiple challenges, including the following:

- Requirements of the material, including environmental sensitivity of medications and equipment

- Quantity and variety of material required, which may vary greatly by disaster type (e.g., PPE and ventilators in a pandemic versus trauma-related needs in an earthquake)

- Lack of medical facility mobility, requiring transportation of supplies to facilities or use of temporary treatment facilities

- Managing surges in demand and logistic resources and providing the elements needed to deliver mass critical care (referred to as surge capacity logistics)

Logistics Operational Challenges in a Pandemic In simplified terms, the two key operational challenges are the movement of medical supplies and equipment into the system and removal of medical waste from the system. The key priorities for influx are the rapid and efficient loading and unloading of supplies. Cargo capacity and the ability to handle the sizes of medical equipment, including standardized disaster response pallets, are critical capabilities. Nondedicated transportation units may have limited capability to be reconfigured for logistical support.

Several innovative logistical transportation solutions were introduced during the COVID-19 pandemic including the possibilities of using eVTOL (e.g., Figure 10) for on-demand medical equipment and supply delivery. In a more traditional approach, M.D. Helicopters, Inc. used a medical helicopter for transporting medical supplies to the Navajo Nation in remote areas of Arizona covering 27,000 square miles. The sheer area, terrain, and obstacles to ground transportation illustrated the potential value in using aerial assets for the rapid delivery of critical medical aid during a pandemic [32]. European eVTOL company Volocopter demonstrated an autonomous utility drone capable of carrying payloads of up to 440 lb (200 kg) over a range of 25 miles, while AutoFlight is developing long-endurance UAS with smaller payload capacities.

Additionally, the logistical system must be able to handle the removal of medical waste through reverse logistics [33]. As illustrated by the COVID-19 outbreak in Wuhan, infectious medical waste exponentially increases during a viral pandemic. A key logistical goal is the minimization of long-term storage of medical waste at the source to reduce the risk of the disease spreading among patients, medical staff, and others. An additional related concern is decontamination of all patient and waste transportation platforms.

Confronting Pandemic Logistical Challenges Several logistical general recommendations and best practices are suggested by the recent COVID-19 pandemic. First, the use of all available transportation assets, including dual or multipurpose use, should be prioritized in a pandemic, including future eVTOL platforms. In particular, for eVTOL platforms, the dynamic allocation of these resources for both patient transportation and logistical support can significantly increase the surge capacity of the

FIGURE 10. eVTOL systems provide lower-cost transportation for people and cargo with greater flexibility in take-off and landing locations in emergencies.

Reprinted with permission. © Johnny T. Doo

transportation system to respond to a pandemic. As illustrated earlier, rapid, flexible repurposing of cost-effective eVTOL vehicles represents a significant opportunity for medical and logistical support in large-scale disasters.

For supply logistics, the modular design of push packages, boxes, or pallets should be considered and matched to the best available platform for transportation. Additionally, strong consideration should be given to potential multirole use of transportation assets during the design phase (e.g., transporting patients, medical supplies, and medical equipment), especially in the novel eVTOL industry. Finally, reverse logistics must also be anticipated. Potential eVTOL solutions include modular vehicle design and rapid vehicle reconfiguration to allow for ease of rapid decontamination after transportation of infectious waste and multirole applications.

Finally, in public health emergencies, an effective medical logistic response requires close coordination between public health agencies and supporting logistical systems to forecast demand and respond dynamically with seamless, timely delivery of medical relief including personnel and supplies. In support of a logistical support system, near-future smart eVTOL technology and command and control systems should prioritize interconnected, interoperable, semi- or fully autonomous systems that anticipate and proactively respond to rapidly shifting medical and logistical transportation needs.

eVTOL Capabilities for Medical Transport and Logistic Applications

The unique features and characteristics of eVTOL aircraft offer flexible deployment, on-location pickup, hospital transfers, as well as suburban and intracity transportation capabilities for medical transport and logistic applications. The nature of most eVTOL designs offer reduced mechanical complexity, lower maintenance requirements, and more cost-effective operations when compared to helicopters or tilt rotors [34]. The design of multirotor, electric power lift systems provides fail-safe features and enhanced reliability in many of the designs, a highly desired parameter for medical transport and logistic missions in critical times. As a necessity, multirotor eVTOL aircraft mostly use a fly-by-wire system to control the motors and rotors, resulting in much simpler pilot/operator control inputs and enhanced vehicle stability (which provides better ride quality in turbulence). The autonomous and semi-autonomous feature of many eVTOL designs allows flexible and dynamic deployment options that may not be suitable for conventional vertical-lift flight vehicles.

Flexible Deployment

Typical helicopters often have specific landing and take-off field requirements, which typically limits them to take-off and land at a few select rooftops and hospitals. Many of the new eVTOL concepts have a smaller landing footprint requirement, allowing for operation from many unconventional locations. The nature of the flight computer that stabilizes rotorcraft including multirotor eVTOL designs enhances the ability of the system to access many additional sites, including open parking lots, schoolyards, urban open fields, etc. This site flexibility makes eVTOL systems much more adaptable for medical patient and equipment/supply transport. However, there are limitations to some designs. The cruise-focused eVTOL has the benefit of extended range and efficiency, but many have a wingspan that might limit landing at certain sites. Since missions may require operations spanning urban, suburban, and rural areas, and these missions may have different transport requirements that vary by region or country, offering a range of eVTOL capabilities in conjunction with ground vehicle operations provides a comprehensive transportation strategy. Additionally, lower eVTOL noise levels may allow operations in many places that might have a "community resistance" to typical emergency transportation during pandemic situations.

On-location Pickup

Significant UAM investment, especially in air taxi vehicles and operations, has led to the development of large business models based on networked operations from selected pre-built vertiports. An urban/suburban vertiport network has the potential to place an air transport pickup point near the patient and a drop-off point near a major hospital center. Additionally, potential markets exist for personal on-demand, to-location use of eVTOL platforms. While landing sites for personal eVTOL aircraft present challenges (i.e., finding landing sites other than the owner's backyard or GA airfield), future expanded use of personal eVTOL platforms may introduce new opportunities for additional unconventional landing sites. Both air taxi services and personal eVTOL platform use, when realized, may provide many readily available landing sites or open spaces for operations, permitting much-needed ability to deploy medical eVTOL into urban, suburban, and rural areas. During a pandemic, availability of multiple eVTOL sites for on-location or near-location patient pickup will significantly reduce transport time to medical centers while simultaneously reducing the need for multimodal transport (and potential exposure to an infectious patient). The time to care in a near-direct flight to destination can be a significant benefit for patient treatment. Ideally, a system linking awareness of the medical care network's capabilities (i.e., specialist, hospital, bed availability) should be established and integrated into the command and control system for the air/ground transportation network to leverage both local and regional medical facility resources. The same smart control system would further offer the capability to allocate medical equipment and supplies as infection cases rise and care needs evolve both within and between areas, as flexible on-demand allocation of resources is critical to battling a pandemic on a day-to-day basis.

FIGURE 11. Sabrewing has secured a two-year contract from the US Air Force to demonstrate the technology in its Rhaegal cargo drone under the Agility Prime program.

Reprinted with permission. © Sabrewing Aircraft

Hospital and Local Transfers

As learned from the COVID-19 case, during a pandemic, it is likely that hospitals will be overwhelmed in a specific metropolitan area, even when other local area hospitals may have available capacity and equipment. Additionally, some hospitals are better able to manage more serious patients (e.g., large urban hospital systems) while other facilities are significantly less capable (e.g., smaller rural hospitals). Autonomous, semiautonomous, or remotely piloted eVTOL platforms that are designed to operate in a complex flight environment should be utilized for on-demand interfacility patient transfer. Most hospitals have rooftop or ground heliports that can be utilized for eVTOL operations, while others can make use of parking lots for this purpose. It would be necessary to provide fueling or charging functions for interfacility eVTOL transfer operations, therefore permanent, temporary, and/or mobile charging/fueling stations (perhaps managed and operated by the Federal Emergency Management Agency or activated National Guard units) should be made available to provide operational support in a pandemic. As developing autonomous transportation systems (both air and ground) is evolutionary and rapidly advancing, the development and deployment of more advanced systems (ideally fully autonomous vehicles) must continue. In parallel, existing systems must be leveraged with the best available levels of autonomy to minimize the infectious risk in pandemic transfer operations.

Intercity, Suburban, and Rural Transport

As a pandemic is not likely to be limited to only a specific urban area or even major cities, the ability to manage intercity, suburban, and rural patient transport and supply logistics is essential. Longer-distance equipment and supply transportation and re-distribution are critical functions. Fortunately, some of the eVTOL designs enable long-range (1,000 nautical miles, Figure 11) intercity people transport and cargo logistics [35]. While some eVTOL aircraft are capable of operating from existing hospital heliports, thousands of GA airports— located in many small towns, suburbs, and rural areas— can serve as operations hubs. These airports could provide refueling support to long-range eVTOL aircraft with hybrid power systems. The GA airport usage model can be especially productive in the United States and its territories as there are over 19,000 airports, heliports, seaplane bases, and other landing facilities [36]. In addition to the 378 primary airports that support scheduled commercial air service, the US also relies on other 2,952 landing facilities (2,903 airports, 10 heliports, and 39 seaplane bases) to support aeromedical flights, aerial firefighting, law enforcement, disaster relief, and remote community access. Additional logistics from these landing sites can be provided via short-range eVTOL or ground transportation to the final destinations.

As experienced during the COVID-19 pandemic, one of the key challenges is the lack of adequate essential medical equipment (e.g., ventilators and PPE). While countries gear up production of such material and supplies, the use of on-demand logistics can optimize equipment transport and deployment. One version of this might be depot stationing of equipment and supplies at various aviation hubs to allow for dynamic transport and delivery to highly infected areas and medical facilities. The mix of longer-range eVTOL systems with short-range urban eVTOL system can be a powerful logistic network for pandemic operations.

Challenges and Development Needed

As a broad array of requirements and missions exist in supporting global pandemic operations, many types of eVTOL design and capabilities are needed. A large number of eVTOL platforms are currently under development, mainly falling into several use categories: intercity air taxi, recreational, cargo delivery, and intracity transport.

Vehicle-Specific Capabilities

Modular Configuration Many eVTOL designs are optimized for a specific mission (e.g., air taxi, cargo delivery, and personal recreation, Figure 12). However, a pandemic response may require these vehicles to be able to convert into a patient transport platform [37] (potentially with a PIU) or into a short- or medium-range medical supply delivery vehicle. The provision for conversion needs to be built into many of these platforms without significant weight and cost penalty, and specific standards need to be developed so that the industry can make these vehicles with a standard interface for a medical stretcher, PIU, cargo restraint, and other features.

Transportability Battery-based eVTOL aircraft tend to have limited range due to the current (and near-future) battery technology and achievable real-world energy density. These eVTOL aircraft may need to be transported on the roadways (Figure 13) to major pandemic sites, from hot zone to hot zone, or even from region to region for rapid and dynamic deployment/redeployment without time-consuming partial or complete disassembly. This transportability capability would ideally be built in during the early basic design phase. This capability would also be critical to transporting in eVTOL aircraft in cargo aircraft (e.g., a military C-130 or commercial cargo aircraft) in support of regional or global redeployment.

Payload Range and Endurance Many eVTOL aircraft currently under development have ranges limited to 100 miles or less due to the constraints of current aviation-safe battery technology. Some designs leverage hybrid or alternative energy power systems to extend useful payload range. Hybrid-powered eVTOL platforms may be able to leverage the existing diesel/jet fuel supply infrastructure or military logistic system, and their extended range may be instrumental in supporting the pandemic mission, especially with mid-range logistics. A mission-driven evaluation of pandemic response requirements can help establish the framework for effective short- to long-range eVTOL mixed capabilities and charging/fuel supply logistic requirements.

FIGURE 12. Flexible interior design and installation can be valuable in converting to patient and supply transport.

Reprinted with permission. © Johnny T. Doo

FIGURE 13. Roadway transportability can be of great value in case of the pandemic situations requiring dynamic redeployments.

Reprinted with permission. © Johnny T. Doo

Take-off and Landing Field Size Requirement To support a broad range of use cases in a pandemic, eVTOL platforms with smaller overall take-off and landing footprints lend to flexible deployment. The overall vehicle size is determined by the basic aircraft configuration of each eVTOL platform, but the field size requirement can be an important design consideration during development. At times, a smaller payload eVTOL platform that can carry one or two people may be more flexible in a minimally supported mission deployment, while a larger unit may be used when adequate support facilities are available and a higher volume of transportation capacity is required. For the same payload capacity, a smaller footprint eVTOL has deployment benefits, including greater flexibility in a dynamic situation involving uncertain or varying take-off and landing locations.

Fast-Charging Relay Stations One option in transferring an eVTOL fleet to a deployment site exceeding the vehicles' range limitations would be to set up fast-charging relay stations along the flight corridor, a necessity for all-electric battery-powered eVTOL platforms. This technology would also be needed and developed for regular air taxi and other eVTOL operations [38]. These mobile fast-charging stations might be towed to open field locations (i.e., serving like large truck stops along the highway). In addition, a study should be performed to determine if an array of automotive supercharger stations (Figure 14) could be combined to provide enough power to support eVTOL relay deployment [39]. The development of 1-3 Megawatt eTruck charging standards and systems make the longer-term outlook even more promising. These approaches and systems would provide a flexible range-extension support infrastructure for the dynamic situations expected during a pandemic.

FIGURE 14. An array of Tesla supercharger stations.

JL IMAGES/Shutterstock.com

Autonomy, Training, and Air Space Management

Semi- to Fully Autonomous Capabilities Most of the eVTOL aircraft are designed to have a certain level of semiautonomous capability for reduced workload, increased safety, and lower operational cost [40]. The road to full autonomy is not just onboard sensors and the processors, but also the ability to handle unplanned events and scenarios that may incorporate artificial intelligence (AI) functions. For pandemic-support missions, the flight route and operating environment are likely different from day-to-day operations. It would be necessary for the semi- or fully autonomous system to be able to handle unplanned and unfamiliar situations. Further development and testing of both self-contained airborne autonomous system and transportable ground-based support units are needed to ensure robust and flexible operations in this kind of dynamic environment.

SVO and Training Most eVTOL utilize indirect flight control systems (often called "fly-by-wire") where the "pilot" provides input to the flight computer, instructing what they want the aircraft to do at a command level. There is no direct "pilot-in-command" control of the surfaces or motor speeds. This indirect control is similar to control method used to pilot enthusiast "drones." It can require less training compared to flying a traditional helicopter or a fixed-wing aircraft because the flight computer handles many of the functions to keep the vehicle stable and perform as requested.

The SVO construct specifies key principles on top of an indirect flight control system. In particular, SVO implies fail-functional systems design and no reversionary operating modes. If an indirect flight control system follows these design tenets, it has the potential to simultaneously reduce the training burden while increasing the level of safety.

Traditional cockpit automation reverts low-level control back to the pilot when the system fails, thereby necessitating that pilots are trained and proficient at performing the underlying function. In the SVO construct, there is no reversionary "direct pilot control" mode allowed—the system must continue to function normally in the presence of all failures. This SVO construct allows the system to be relied upon to replace a large portion of pilot training. There are existing SVO constructs in the field today (e.g., a full-authority digital engine control or "FADEC").

This simplified control enables accelerated training for a sufficient number of eVTOL pilots/operators. NASA has been working on eVTOL training systems to evaluate the human response to the flight training and pilot behavior to best design the cockpit and/or remote-control station for eVTOL aircraft (Figure 15).

The GAMA Electric Propulsion & Innovation Committee "EPIC" SVO Subcommittee published a white paper in May 2019 that addresses the use of automation to reduce pilot/operator training while increasing the level of safety [41]. The paper connects aircraft certification requirements to pilot/operator training requirements.

The SVO concept is critical to mass deployment of eVTOL aircraft into commercial operations, and it is even more important when it comes to handling the global pandemic in that additional pilot/operators may be needed in short order to provide the dynamic deployment required to deal with increased demand. The possibility of training medical workers to "operate" an eVTOL platform during a pandemic is highly attractive. It further highlights the importance of automation of eVTOL technology, in that these individuals can be activated as reserve forces as the need arises.

Dynamic Airspace Management It is expected that a functional airspace management system will be developed in time to support and manage forthcoming UAM operations [42]. The NASA AAM initiative, in conjunction with the FAA,

FIGURE 15. eVTOL simulation training system at NASA Ames Research Center.

Reprinted with permission. © Johnny T. Doo

FIGURE 16. Even in high-density megacities in Asia, rooftop heliports and school fields can be used as temporary vertiports to support eVTOL operations.

is addressing this aspect not only to cover the eVTOL operations, but to integrate with traditional aviation and rapidly expanding UAS activities. What needs to be addressed is that the system must be configured such that it can handle normal day-to-day operations and be able to reconfigure dynamically and manage unforeseen pandemic mission requirements and activities.

Operations, Logistics, and Support

Fast Deployment and Redeployment Fleet Management An emergency response fleet management program would need to be established, maintained, and exercised periodically. It would need to be integrated between federal, state, and local governments similar to other emergency management protocol. The program would be applicable for various emergency situations, including hurricanes, earthquakes, tsunami, and other disasters. One key difference would be that the management plan needs to be able to not only quickly deploy to the emergency locations (like other natural disasters), but also dynamically redeploy to other areas as needed during the course of the emergency (e.g., pandemic).

Temporary Vertiport Setup and Management Given the difficulty with predicting when, where, and to what degree each area or region may be affected by a "black swan" event such as a pandemic, an "antifragile"

approach would focus on developing capabilities to allow flexible, on-demand vertiport basing for the eVTOL fleet. One potential method would be to have available support units that can travel via roadways to create temporary vertiports by leveraging existing open areas like shopping malls or hospital parking lots, schoolyards, and even open grassy fields (Figure 16). One vertiport setup that can be adaptable early is the use of the GA airport; many of these have the necessary airspace and ground operations or can quickly be adapted to support eVTOL emergency operations (although fast-charging stations must be considered). This is especially workable in the US and somewhat in Europe, but may be more limited in many other countries or regions [43].

Ground Support, Spare Parts Logistics, and Maintenance As mentioned, (at the minimum) flexible eVTOL operations will need fast-charging stations or an appropriate fuel supply to support electric or hybrid eVTOL energy requirements. Additionally, the trained supporting ground crew will have to be available on short notice. As for any vehicle, especially flying ones, controlled spare parts must be available at the depot level and shippable to the user site as required. Therefore, a logistics plan would be required long before any actual pandemic or emergency could be supported. One key benefit of using the eVTOL fleet is that it can likely serve as a spare-part delivery service without another logistics system, but such use will require planning and management.

Distributed On-demand Network System It is likely that the high volume of the transportation sorties needed to support a pandemic situation will overwhelm any dedicated emergency response fleet. Because of the special functions and configurations required for emergency response missions, a purpose-designed and -built first-response fleet would be required regardless. But it would also be necessary to establish a practice and protocol to activate (draft) a certain percentage of the civilian fleet (like air taxis) to sustain pandemic operations. The benefit is that the local emergency responder team would be the first to be activated, but the civil team can be added quickly afterward. When local conditions are more under control, the drafted fleet can then be released back to normal services. In addition, a set of standards would need to be established so that common command, control, and communication systems are built into all civilian eVTOL platforms so that they can be managed effectively and safely when activated.

Resilience of eVTOL Operational Chain

The COVID-19 pandemic often highlighted a lack of preparedness. This can be seen in the outsourced production capacity and insufficient stockpiles of PPE that put nursing-home residents, community healthcare workers, and hospital staff at risk, weakening healthcare systems further. An eVTOL supply chain capable of handling the dynamic characteristics of logistic models under disruption—and their recovery processes—is expected to be developed within the future eVTOL vehicle introduction. For this, the concept of resilience is critically important for implementing eVTOL platforms into emergency response. Resilience, defined as the ability to recover quickly and effectively from a disruption, is a key, multifaceted concept in risk management. The crucial idea is that nothing can be anticipated in every detail, and thus, what the system needs is to be "prepared to be unprepared." Achieving eVTOL resilience in response to a pandemic requires the incorporation of essential system characteristics such as robustness, stability, diversity, redundancy, flexibility, resourcefulness, coordination capacity, modularity, collaboration, agility, efficiency, creativity, equity, foresight capacity, self-organization, and adaptability of the eVTOL operational chain to the pandemic characteristics.

Unconventional Medical Transport Mode Evaluation

Transport from Contaminated Site to Hospital or Care Center

Regarding medical transport, one case is of particular interest: moving people from a cruise ship to quarantine locations or hospitals (Figure 17). During the COVID-19 response, many medical and support personnel were on-site to manage the situation. In the future, that amount of personnel could be reduced if autonomous or remotely operated eVTOL platforms are available to transport patients with minimal ground crew and no pilots. It may take some years to establish the functionality and reliability for such operations, but it would be beneficial in an emergency or pandemic environment.

On-demand Distribution of Medical Equipment and Supplies

When used for cargo transportation, eVTOL platforms can be managed similar to UAS, even using the UAS traffic management system developed by NASA and the FAA or similar systems in other countries [44]. In this mode, even smaller eVTOL like personal/recreational units could be used to provide more location flexibilities in conjunction with larger capacity units. The key would be to have packaging strap-down and cushioning provisions and to train nonprofessional eVTOL ground crew to support such operations. A decontamination process should be established analogous to patient transportation missions. The flight operations could be interlinked with ground transportation to minimize air traffic due to the use of passenger or patient transport eVTOL. Fortunately, ground traffic could be minimum, especially during lockdown. A ground-air-ground air taxi operating model and reservation system could also be modified to support pandemic logistic functions.

Medical Personnel and First-Responder Dynamic Allocation

In some cases, the infected population that requires hospitalization can shift from one zone to another,

---

FIGURE 17. A cruise ship with helipad could transfer patients directly to the hospital with eVTOL.

even among cities and suburban areas. To handle the shortage of medical personnel, an on-demand transportation system would allow better allocation of human resources to maximize effectiveness in a pandemic. Perhaps instead of developing a new flight management system, the existing system supporting the UAM system could be used for medical and support dynamic personnel allocation. What needs to be developed and implemented is the emergency personnel database and demand management system that leverages the eVTOL capability to complement other transportation options.

Longer-Range Regional Logistics In addition to local transportation logistics operations, optimized long-range regional transportation capabilities to manage scarce medical equipment and supplies are needed. There are several eVTOL developers working on long-range, hybrid-winged eVTOL platforms that can provide an operating range of up to 500 nautical miles. This type of vehicle can provide the needed logistic capability to form an integral cargo shipment and delivery system with other short-range eVTOL and also transport patients or medical personnel to support other hot spots.

Funding and Acceptance

Federal, State, and Local Government Funding and Support In the situation of a global pandemic, the federal, state, and local governments would likely allocate a significant amount of money to support the effort of fighting the rapidly transmitting diseases. There should be two parts of the funding plan under consideration: one is the funding needed at peacetime to prepare and set up the specialized emergency response eVTOL fleet, and the other would be the required funding to activate the general-purpose fleet, support personnel, operating expenses, and industry reimbursement once an emergency starts. It is imperative to establish a fleet of eVTOL aircraft designed for medical and cargo transport. This fleet should be assigned to an operating

group (e.g., the US Army National Guard) that can support public service missions and be available to support emergency operations [45]. Funding for an emergency response fleet would need to be ongoing. The second part concerns general-purpose eVTOL, like air taxis and cargo transports, that can be drafted during a pandemic under unified coordination and command. The funding required for the drafting, operation, and support of the general-purpose fleet would come from emergency response funds.

User/Passenger Acceptance At the earlier stage of the eVTOL market development, not everyone in the general public had the chance of flying in one. The concern of reliability and safety may vary for the first responders, medical personnel, and patients. Therefore, an early adoption of eVTOL technology for public service missions would be essential to establish the "for Good" mission feasibility and experiences and build up the trust with the general public. Early success missions with lives saved would significantly enhance acceptance during a pandemic. After eVTOL applications become widespread (e.g., for air taxis, personal use, and public services), the concern and acceptance would likely not be an issue anymore.

General Public and Community Support While it might be expected that the public would support the use of eVTOL to help transport patients and medical supplies during a pandemic, there can be significantly different public opinions regarding the best way to fight diseases such as COVID-19 [46]. It is therefore vital to engage the communities early and communicate that eVTOL platform use for public services can help save the lives of people in our own communities.

Global Challenges Not all countries and regions will be adopting eVTOL fleets for special missions and general-purpose operations at the same pace. Yet, the benefit of leveraging eVTOL technology can make a significant difference in dealing with a pandemic (Figure 18). Ideally, an eVTOL fleet

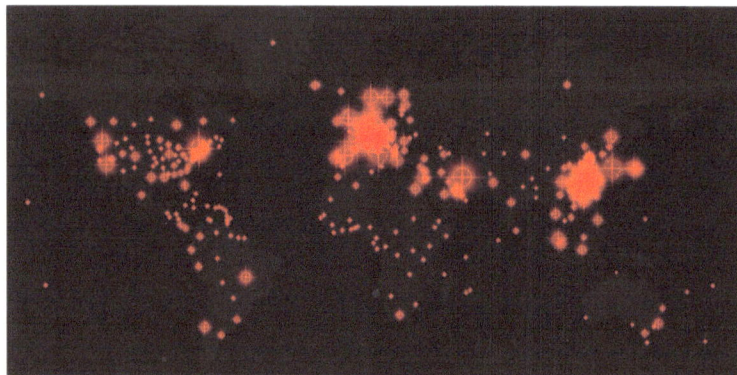

FIGURE 18. **The global nature of a pandemic.**

Sebestyen Balint/Shutterstock.com

and its operating capabilities could be transferred to another region that is impacted more severely. Transferring assets is a significant logistic challenge, and a predetermined plan must be established and practiced. Specific types of eVTOL platforms are more suitable to be transported, especially ones with a smaller packing/shipping size and less required support/maintenance. Long-range cargo transport and personnel delivery platforms would naturally be the first wave of support. The shorter-range fleet would likely have to be transported via air. Interregional support would be hard to coordinate, but the only way it will be successful is to evaluate the feasibility and establish a contingency plan early on.

Social Benefits

Life-Saving

eVTOL technology offers significant gains in total transportation speed compared to the current ground or air medical transportation assets. These include shorter time from activation to response compared to HAA, faster straight-line travel time than GEMS, and greater safety than current HAA (through SVO and semi- or fully autonomous vehicle control).

Medical Care Effectiveness Increase and Transport Cost Reduction

Through simplified vehicle design and operations, eVTOL technology offers significant cost-savings compared to current HAA. While HAA has medical life-saving advantages over GEMS, the current use of HAA is greatly limited by the cost of operations, leading to cost-based decision-making. However, with the potential for significant savings with an eVTOL EMS, the decision to use such platforms for an expanded set of rescue and patient transportation situations may afford more rapid transportation for a greater number of patients. With cost becoming less of a barrier to rapid transportation, this life-saving technology may translate into greater population-level benefits.

Medical Supply and Equipment On-demand Logistics

As discussed, a smart network of eVTOL vehicles capable of flexible, modular use may offer the medical logistics sector significant benefits. Through smart command and control of a distributed network of semi- or fully autonomous eVTOL platforms, logistic networks will become more resilient, with

rapid and robust operations. With such a network, real-time/on-demand logistics becomes possible. Ultimately, near-future, AI-enhanced command and control systems that are proactive and predictive may enable anticipating and addressing logistical challenges before they become rate-limiting to medical operations.

Enhanced Society Acceptance of eVTOL Technology

eVTOL platforms represents a convergence of key technologies, including electric power and intelligent computerized control, that enable novel capabilities. The transformative capabilities of eVTOL technology unlock new roles and possibilities, including aeromedical and logistical operations, with higher levels of safety and cost-savings than present transportation options. In addition to medical and logistical operations, eVTOL platforms will also be concurrently entering the public space as air taxis and delivery "drones." As members of the public encounter eVTOL vehicles in an expanding number of industries and applications, society will experience the numerous benefits provided by eVTOL, many of which are truly transformative. Similar to the adoption of other radically transformative platforms (such as the internet and the smartphone), society will reach an inflection point, entering a new era in which it will wonder how it was able to live without it.

Summary

Need for Partnerships and Consolidated Implementation Approaches

Leveraging the eVTOL industry's rapid development and future capacity for medical transport and logistic capabilities in response to a global pandemic will require a concerted team effort to accomplish. The total eVTOL operational capability is significantly greater than simply a new generation of vertical take-off and landing aircraft. It includes and requires the operating system, government and community engagement, airspace control and management, supporting services, on-demand operating fields, and distributed maintenance and fueling/charging capabilities.

These efforts cannot be engaged individually—they will be accomplished by partnerships and consolidated implementation approaches. The ability to leverage collaborative industry standards and best practices would minimize duplicated efforts, especially the numerous functions and complex operations required for an effective, on-demand pandemic response.

The global nature of a pandemic further emphasizes the need for cooperation and collaboration in this quest. A concerted effort within a country is one thing, but a worldwide partnership in the development and deployment of the eVTOL fleet and command and control systems would greatly accelerate effectiveness. Cross-border and intercontinental logistic support would likely be needed for emergency and pandemic response.

SAE EDGE™ Research Reports

SAE EDGE™ Research Reports, like the present report on "Unsettled Issues Concerning the Opportunities and Challenges of eVTOL Applications During a Global Pandemic," are intended to push further out into still unsettled areas of technology of interest to the mobility industry. SAE launches these reports before attempting to form a joint working group, let alone a cooperative research program or a standards committee.

SAE EDGE™ reports are intended to be quick, concise overviews of major unsettled areas where vital new technologies are emerging. An unsettled area is characterized more by confusion and controversy than established order. Early practitioners must confront an absence of agreement. Their challenge is often not to seize the high ground but to find common ground. These scouting reports from the frontiers of investigation are intended merely to begin the process of sorting through critical issues, contributing to a better understanding of key problems, and providing helpful suggestions about possible next steps and avenues of investigation.

SAE EDGE™ Research Reports, therefore, are fundamentally distinct from the more formal working groups approach and far removed from the more mature research program and standard's development process.

Next Steps for eVTOL Vehicles and System Development

To effectively leverage eVTOL systems for pandemic transportation and logistic support, capabilities and operations must be planned for and executed beforehand—not as a reactive solution when the next pandemic hits. Based on the rapid development momentum of vehicles and operating system, it is envisioned that there will be a sizable fleet in operation with many different designs and configurations available for various missions, possibly within the next few years. While SAE International and other industry organizations have done extensive work related to eVTOL design, certification, and operating standards, the evolving nature of this area requires continuous development to keep pace with changing industry trends and technological innovations. This SAE EDGE Research Report on the opportunities

and challenges of eVTOL applications in a global pandemic identifies the following critical issues for further discussions among industry stakeholders, specific working groups, and dedicated standardization committees:

- **The need for unified command and control system standards:** Evaluate the possibilities to embed a set of standard command and control protocols that can be activated in the time of emergency response and services for more standardized vehicle management.

- **The need for modular payload specifications and interface standards for quick cargo conversion and medical transport conversion:** When needed, this will allow for a depot inventory of mission modules and quarantine chambers for pandemic operations.

- **The need for a flexible airspace management system leveraging the AAM airspace management system currently under development:** This also accounts for high-volume, dynamic eVTOL deployment during emergency operations with rapidly shifting missions and flight plans beyond normal vertiports.

- **The need for mobile fast-charging/fueling station standards:** The development of fast-charging/fueling points will enable flexible, on-demand deployment capabilities, and establishing relay stations at unprepared locations supporting the eVTOL fleet migration as needed.

- **The need for evaluating the feasibility and possible methods to leverage automotive fast-charging stations:** The power source for emergency eVTOL transfer charging methods is of vital importance for rapid and diverse deployment of eVTOL assets, especially when operated in an emergency or pandemic situation.

Recommendations

The overall recommendations of this SAE EDGE™ report can be summarized as follows:

1. The eVTOL and supporting industry should consider the benefit and potential use of eVTOL platforms to make a significant contribution in a global pandemic response.

2. The industry should take design features and operational requirements for possible emergency use. Tailor the vehicle and system design with configurations for quick adaptability for pandemic missions.

3. Leverage the ongoing rapid technology and platform development as the base for the pandemic mission needs. Identify and collaborate specific technologies including sense-and-avoid capabilities,

flexible airspace management, and flexible vehicle configuration for medical and cargo missions with rapid decontamination and quarantine features.

4. Unify communication and control standards in a coordinated manner, leveraging professional organizations such as SAE International, ASTM, and GAMA to ensure interoperability among systems to enable mobilized fleet-compatible command, control, and communication.

5. Government agencies, business entities, vehicle and system manufacturers, and operators should all be part of the on-demand logistic network that can be activated at short notice where and when they are needed. This on-demand ability is by far the most important factor regarding eVTOL technology in support of global pandemic operations.

6. Collaborate with government emergency response and logistic agencies, hospitals and medical care organizations, eVTOL operators, local and regional communities, as well as aviation governing and managing agencies. Form a joint force team to plan and organize the potential operational needs and set an executable plan to respond when needed.

The overarching goal of the above recommendations is to encourage and enable eVTOL technology development and system integration, leading to industrial-level fleet deployment and operations, and enabling mission-specific systems and on-demand adaptation of available fleets worldwide to enhance emergency management capabilities.

Abbreviations/Definitions

AAM - [NASA] Advanced Air Mobility

AI - Artificial Intelligence

ATP - Autonomous Transport Pod

CFR - Code of Federal Regulations

ConOps - Concept of Operations

COVID-19 - Coronavirus Disease 2019

EMS - Emergency Medical Services

EMT - Emergency Medical Technician

eVTOL - Electric Vertical Take-Off and Landing

FAA - Federal Aviation Administration

GA - General Aviation

GEMS - Ground Emergency Medical Services

HAA - Helicopter Air Ambulance

HEMS - Helicopter Emergency Medical Services

IFT - Interfacility Transfer

lb - pound

MEDEVAC - Medical Evacuation

MRO - Maintenance, Repair, and Overhaul

mph - miles per hour

NASA - National Aeronautics and Space Administration

NPRM - Notice of Proposed Rulemaking

NTSB - National Transportation Safety Board

OEM - Original Equipment Manufacturer

PIU - Patient Isolation Unit

PPE - Personal Protective Equipment

sUAS - Small Unmanned Aircraft System(s)

SVO - Simplified Vehicle Operation

TRL - [NASA] Technical Readiness Level

UAM - Urban Air Mobility

UAS - Unmanned Aircraft System(s)

UAV - Unmanned Aerial Vehicle

US - United States

Acknowledgements

Recognition should go first to all the participants, many of whom also provided feedback on the draft version of this publication. Without their input and initiative, this SAE EDGE™ Research Report would not have been possible.

Christopher M. Edens, MD, US Air Force
Michael R. Smith, Bell Textron, Inc.
Marilena D. Pavel, Ph.D, Delft University of Technology
Carl Dietrich, Jump Aero, Inc.
Ed De Reyes, Sabrewing Aircraft Co.

Our gratitude goes to the NASA Transformative Vertical Flight Working Group 4—Public Services for inspiring discussions on the future of eVTOL aircraft; the US Air Force Agility Prime community for their enthusiasm on ushering the next era of aerial vehicles; the Vertical Flight Society for their continuous support to advance eVTOL aircraft. Gratitude for the heroes of healthcare and first responders, who put their own safety at risk.

The authors of this document together with the SAE Team responsible for its creation join in expressing our deepest appreciation to all the individuals mentioned above.

Johnny T. Doo
International Vehicle Research, Inc.

References

1. Vertical Flight Society Electric VTOL News™, "Supporting the Electric VTOL Revolution, eVTOL. news Hits 300," Vertiflite Sep./Oct. 2020, http://evtol.news/.

2. NASA, "Advanced Air Mobility (AAM)," https://www.nasa.gov/aam, accessed June 29, 2020.

3. "Agility Prime," https://agilityprime.com/, accessed June 29, 2020.

4. NASA, "Transformative Vertical Flight Working Groups," https://nari.arc.nasa.gov/wghome, accessed June 29, 2020.

5. Worldometer, "COVID-19 Coronavirus Pandemic," 2020, https://www.worldometers.info/coronavirus/, accessed Oct. 15, 2020.

6. Caduff, C., "What Went Wrong: Corona and the World after the Full Stop," *Medical Anthropology Quarterly*, May 2020.

7. McKinsey and Company, "COVID-19: Briefing Materials. Global Health and Crisis Response," https://nicfraternity.org/mckinsey-company-covid-19-briefing-materials/, accessed June 2020.

8. Vertical Flight Society, "EHang Demonstrates Medical Transport Applications in Coronavirus Response," https://evtol.com/news/ehang-medical-transport-coronavirus, accessed Feb. 25, 2020.

9. Choi, T.-M., "Innovative "Bring-Service-Near-Your-Home" Operations under Corona-Virus (COVID-19/SARS-CoV-2) Outbreak: Can Logistics Become the Messiah?" *Transportation Research Part E* 140:101961, Apr. 2020, https://doi.org/10.1016/j.tre.2020.101961.

10. Department of Transportation Federal Aviation Administration 14 CFR Part 107, Aviation Rulemaking Committee (ARC), NPRM 84 FR 3865, Feb. 13, 2019.

11. Federal Aviation Administration, "UAS by the Numbers," https://www.faa.gov/uas/resources/by_the_numbers/, accessed Oct. 15, 2020.

12. NASA, "Definition Of Technology Readiness Levels," https://esto.nasa.gov/files/trl_definitions.pdf, accessed Oct. 15, 2020.

13. NASA, "Advanced Air Mobility," Feb. 26, 2020, https://www.nasa.gov/aeroresearch/programs/iasp/aam, accessed June 28, 2020.

14. U.S. Air Force, "Thinking Ahead: Agility Prime Kicks Off," https://www.af.mil/News/Article-Display/Article/2172755/thinking-ahead-agility-prime-kicks-off/, accessed June 28, 2020.

15. Reimer, A.P., and Hobensack, M., "Establishing Transport Statistics: Results From the Medevac Transport Statistics Survey," *Air Med J* 38(3):174-177, 2019, https://doi.org/10.1016/j.amj.2019.03.008.

16. Taylor, B.N., Rasnake, N., McNutt, K., McKnight, C.L. et al., "Rapid Ground Transport of Trauma Patients: A Moderate Distance From Trauma Center Improves Survival," *J Surg Res* 232:318-324, 2018, https://doi.org/10.1016/j.jss.2018.06.055.

17. Ingalls, N., Zonies, D., Bailey, J.A. et al., "A Review of the First 10 Years of Critical Care Aeromedical Transport during Operation Iraqi Freedom and Operation Enduring Freedom: The Importance of Evacuation Timing," *JAMA Surg* 149(8):807-813, 2014, https://doi.org/10.1001/jamasurg.2014.621.

18. Curnin, S., "Large Civilian Air Medical Jets: Implications for Australian Disaster Health," *Air Med J* 31(6):284-288, 2012, https://doi.org/10.1016/j.amj.2012.04.001.

19. Di Rocco, D., Pasquier, M., Albrecht, E., Carron, P.N. et al., "HEMS Inter-Facility Transfer: A Case-Mix Analysis," *BMC Emerg Med* 18(1):13, 2018, https://doi.org/10.1186/s12873-018-0163-8.

20. Stewart, K., Garwe, T., Bhandari, N., Danford, B. et al., "Factors Associated with the Use of Helicopter Inter-Facility Transport of Trauma Patients to Tertiary Trauma Centers within an Organized Rural Trauma System," *Prehosp Emerg Care* 20(5):601-608, 2016, https://doi.org/10.3109/10903127.2016.1149650.

21. Ramadas, R., Hendel, S., and MacKillop, A., "Civilian Aeromedical Retrievals (the Australian Experience)," *BJA Education* 16(6):186-190, June 2016, https://doi.org/10.1093/bjaed/mkv040.

22. Shaw, J.J., Psoinos, C.M., and Santry, H.P., "It's All about Location, Location, Location: A New Perspective on Trauma Transport," *Ann Surg* 263(2):413-418, 2016, https://doi.org/10.1097/SLA.0000000000001265.

23. Madhav, N., Oppenheim, B., Gallivan, M. et al., "Pandemics: Risks, Impacts, and Mitigation," in: Jamison, D.T., Gelband, H., Horton, S. et al., eds., *Disease Control Priorities: Improving Health and Reducing Poverty*, Third Edition (Washington, DC: The International Bank for Reconstruction and Development/The World Bank, 2017), Chapter 17, https://doi.org/10.1596/978-1-4648-0527-1_ch17. Available from https://www.ncbi.nlm.nih.gov/books/NBK525302/.

24. Tippett, V.C., Watt, K., Raven, S.G. et al., "Anticipated Behaviors of Emergency Prehospital Medical Care Providers during an Influenza Pandemic," *Prehosp Disaster Med* 25(1):20-25, 2010, https://doi.org/10.1017/s1049023x00007603.

25. Albrecht, R., Knapp, J., Theiler, L., Eder, M. et al., "Transport of COVID-19 and Other Highly Contagious Patients by Helicopter and Fixed-Wing Air Ambulance: A Narrative Review and Experience of the Swiss Air Rescue Rega," *Scand J Trauma Resusc Emerg Med* 28(1):40, 2020, https://doi.org/10.1186/s13049-020-00734-9.

26. Chen, X., Gestring, M.L., Rosengart, M.R. et al., "Speed Is Not Everything: Identifying Patients Who May Benefit from Helicopter Transport Despite Faster Ground Transport," *J Trauma Acute Care Surg* 84(4):549-557, 2018, https://doi.org/10.1097/TA.0000000000001769.

27. Thomas, S.H., and Blumen, I., "Helicopter Emergency Medical Services Literature 2014 to 2016: Lessons and Perspectives, Part 2-Nontrauma Transports and General Issues," *Air Med J* 37(2):126-130, 2018, https://doi.org/10.1016/j.amj.2017.10.005.

28. Bledsoe, B.E., "Air Medical Helicopter Accidents in the United States: A Five-Year Review," *Prehosp Emerg Care* 7(1):94-98, 2003, https://doi.org/10.1080/10903120390937175.

29. Brown, J.B., Gestring, M.L., Guyette, F.X. et al., "External Validation of the Air Medical Prehospital Triage Score for Identifying Trauma Patients Likely to Benefit from Scene Helicopter Transport," *J Trauma Acute Care Surg* 82(2):270-279, 2017, https://doi.org/10.1097/TA.0000000000001326.

30. Motomura, T., Hirabayashi, A., Matsumoto, H. et al., "Aeromedical Transport Operations Using Helicopters during the 2016 Kumamoto Earthquake in Japan," *J Nippon Med Sch* 85(2):124-130, 2018, https://doi.org/10.1272/jnms.2018_85-19.

31. Sahebi, A., Ghomian, Z., and Sarvar, M., "Helicopter Emergency Medical Services in 2017 Kermanshah Earthquake; a Qualitative Study," *Arch Acad Emerg Med* 7(1):e31, 2019.

32. MD Helicopter, "MD Helicopter Launches Flight Support For Covid-19 Humanitarian Missions," https://www.mdhelicopters.com/md-helicopter-launches-flight-support-for-covid-19-humanitarian-missions.html, accessed June 5, 2020.

33. Yu, H., Sun, X., Solvang, W.D., and Zhao, X., "Reverse Logistics Network Design for Effective Management of Medical Waste in Epidemic Outbreaks: Insights from the Coronavirus Disease 2019 (COVID-19) Outbreak in Wuhan (China)," *Int J Environ Res Public Health* 17(5):1770, 2020, https://doi.org/10.3390/ijerph17051770.

34. ARC, "A Study in Reducing the Cost of Vertical Flight with Electric Propulsion," June 5, 2017, https://arc.aiaa.org/doi/10.2514/6.2017-3442, accessed June 27, 2020.

35. Air Cargo, "Sabrewing Set for Launch of Heavy-Lift Drone Aircraft," Apr. 24, 2020, https://www.aircargonews.net/airlines/freighter-operator/sabrewing-set-for-launch-of-heavy-lift-drone-aircraft/, accessed June 27, 2020.

36. U.S. Department of Transportation Federal Aviation Administration, "General Aviation Airports: A National Asset (May 2012)," May 1, 2012, https://www.faa.gov/airports/planning_capacity/ga_study/media/2012assetreport.pdf, accessed June 27, 2020.

37. Osborne, T., "Speedy Adaptations Help Helicopter to Fight Pandemic," *Aviation Week & Space Technology* Apr. 20-May 3, 2020.

38. NREL, "Commercial Vehicles and Extreme Fast Charging Research Needs Workshop 2019," Aug. 27, 2019, https://www.nrel.gov/transportation/assets/pdfs/extreme-fast-charging-research-needs-workshop-2019-presentations.pdf, accessed June 29, 2020.

39. Wikipedia, "Tesla Supercharger," https://en.wikipedia.org/wiki/Tesla_Supercharger, accessed June 2020.

40. Walthall, R., "Unsettled Topics on the Use of IVHM in the Active Control Loop," https://www.sae.org/publications/technical-papers/content/epr2020011/, accessed June 29, 2020.

41. GAMA EPIC SVO Subcommittee, "A Rational Construct for Simplified Vehicle Operations (SVO)," Whitepaper, Version 1.0, May 20, 2019.

42. NTRS, "Urban Air Mobility Airspace Integration Concepts and Considerations," https://ntrs.nasa.gov/archive/nasa/casi.ntrs.nasa.gov/20180005218.pdf, accessed June 30, 2020.

43. NATA, "Urban Air Mobility: Considerations for Vertiport Operation," https://www.nata.aero/assets/Site_18/files/GIA/NATA%20UAM%20White%20Paper%20-%20FINAL%20cb.pdf, accessed June 30, 2020.

44. Federal Aviation Administration, "Unmanned Aircraft System Traffic Management (UTM)," June 3, 2020, https://www.faa.gov/uas/research_development/traffic_management/, accessed June 30, 2020.

45. Aviation Today, "Air Force to Begin Fly-Off, Certification Campaign for eVTOL Aircraft," Feb. 21, 2020, https://www.aviationtoday.com/2020/02/21/government-broadens-engagement-de-risking-efforts-electric-air-taxi-market/, accessed June 30, 2020.

46. Altran, "To Urban Air Mobility En-Route En-Route to Urban Air Mobility," https://www.altran.com/as-content/uploads/sites/27/2020/03/en-route-to-urban-air-mobility.pdf, accessed June 30, 2020.

Contact Information

EDGEresearch@sae.org.